FREE RULER ON EVERY BOOK!

MURDEROUS M

T0173378

The Brain-Bending Basics

IT'S A THOUGHT!

× = i ♥ MATHS

KJARTAN POSKITT

HELLO

■SCHOLASTIC

www.murderousmaths.co.uk

Scholastic Children's Books,
Euston House, 24 Eversholt Street,
London NW1 1DB, UK

A division of Scholastic Ltd
London ~ New York ~ Toronto ~ Sydney ~ Auckland
Mexico City ~ New Delhi ~ Hong Kong

First published in the UK by Scholastic Ltd, 2014
This edition published by Scholastic Ltd, 2019

Some of the material in this book has previously been published in *Murderous Maths: Guaranteed to Bend Your Brain* (1997), *Murderous Maths: Guaranteed to Mash Your Mind* (1998), *Murderous Maths: Desperate Measures* (2000), *Murderous Maths: Professor Fiendish's Book of Brain Benders* (2002), *Murderous Maths: The Phantom X* (2003), *Murderous Maths: The Fiendish Angletron* (2004), *Murderous Maths: The Perfect Sausage* (2005), *Murderous Maths: Easy Questions, Evil Answers* (2010)

ISBN 978 1407 19712 8

Printed and bound in the UK by CPI Group Ltd, Croydon, CR0 4YY

2 4 6 8 10 9 7 5 3 1

www.scholastic.co.uk

CONTENTS

KJARTAN POSKITT's first jobs included playing pianos very loudly, presenting children's tv, inventing puzzles and writing pantomimes. Maths was his best subject at school, because it was the only one that didn't need good spelling and handwriting. As well as 30 maths books, he has written books about space, magic, codes and pants, and he also writes the Agatha Parrot and Borgon the Axeboy books. His favourite number is 12,988,816 because that's how many ways you can put 32 dominoes on a chessboard (although he didn't count them himself). If he wasn't an author he would like to have been a sound effects man. He has two old pinball tables, seven guitars and lots of dangerous old music synthesisers and he plays all of them… badly!

MURDEROUS MATHS –
ARE YOU KIDDING?

If you think maths is just harmless little things like 2+3 = 5 then you're in for a shock! *The Brain-Bending Basics* includes all sorts of weird and strange stuff! But first, to give you an idea of what's coming, look at this story from the police files...

City: Chicago, Illinois, USA
Place: The state prison
Date: 2 December, 1929
Time: 4:00 a.m.

The dynamite flew in through the bars of the window and rolled across the floor.

"Jeez!" screamed One-Finger Jimmy. "She's doing it. She's getting us out!"

"Grab some cover!" shouted Blade. "That fuse ain't gonna wait!"

The seven men ripped their bunks apart and seconds later were cowering in a corner behind a wall of mattresses.

"Move over, you guys," said the biggest man. "My rear is still exposed."

"What's your problem, Porky?" sneered Weasel. "With the size of that thing, if you sat on a fork at breakfast, you wouldn't shout ouch until dinner."

"Clam up!" ordered Blade. "And keep down."

Each man screwed his eyes tight shut as the spark spat its way down the fuse.

"Coo-ee boys!" sang a voice from the door. "Sleeping tight, I hope?" A large key clanged in the lock and the door swung open. Outside was a smart lady smelling of perfume.

"Dolly Snowlips!" gasped Jimmy. "What are you doing there? You're supposed to be waiting over the wall with the pick-up truck."

"Yeah, but that was *your* plan," replied Dolly. "And face it, you guys ain't too smart."

"We is plenty smart enough, sister," said Chainsaw Charlie.

"Is that a fact?" came the reply. "So how come you got arrested for fighting over a restaurant bill? Couldn't you just divide it by seven?"

"Numbers was working it out," said Jimmy.

"Seven sevens, 49. Seven 49s, three-four-three. Seven three-four-threes, two-four-oh-one…" gibbered the thin man.

"He's just a machine in pants," said Dolly. "Sure, Numbers can do the sums, but he doesn't know what sums to do!"

"He does what I tell him," said Blade. "Remember, inside or out, I'm still the boss."

"Oh really?" said Dolly. "Who's the one hiding in the corner in his underwear?"

"Hey! That's you, boss," said Half-Smile.

"And who's just been in the governor's office getting you all bailed out?"

"We're bailed out?" gasped the mob.

"You mean somebody actually paid money so that we're free to go?" said Blade.

"They sure did," said Dolly. "And that's a far better way of getting you out of here than

explosions, sirens and me hanging around outside freezing my knees off in a stolen truck."

"How could we get bail?" said Jimmy. "We're the evilest, meanest, dirtiest dogs that ever did a dastardly deed."

"Yeah, that's us," the rest all chorused.

"That's why bail was ten million bucks," said Dolly.

There was a shocked silence.

"And where did these ten fat ones come from?" asked Blade.

"A friend," said Dolly. "A friend who wants it paid back."

"Where are we supposed to find that sort of dough?"

"The Fort Knocks Wages Express," said Dolly.

"You're joking," gasped Blade. "Nobody knocks off the Knocks Express."

"I got it all figured out," said Dolly.

"Wow!" said Jimmy. "What a dame!"

"Maybe I don't like it," said Blade.

"Maybe no one's asking you," said Chainsaw. "Looks like Doll's the boss now."

"Good," said Dolly. "Well what are you waiting for? Come on."

She turned and set off down the corridor. One by one the bewildered men stepped out to follow her, with the big man at the rear. It was just as he was squeezing himself through the narrow door that he called out: "Hey Dolly! If you got us bail, and you ain't waiting outside with the truck, why did you chuck the dynamite?"

"I didn't chuck no dynamite," replied Dolly.

"Well somebody chucked the dynamite!" said Porky.

Somewhere inside the package on the cell floor, the spark reached the end of its journey.

And as Blade and the gang stumble blinking and coughing out into the street, we've just got one thing to say…

Welcome to the weird world of Murderous Maths!

Oh yes indeed, getting out of jail was the easy bit. As Blade will soon find out, there is no escape from Murderous Maths. It's everywhere and it affects all of us and everything in the whole universe! But don't panic.

Murderous Maths isn't just dull old sums. It's about slick tricks and gruesome games, it's about strange people, it's about taking control of everything that goes on around us…

IT'S ABOUT TIME WE GOT ON WITH THE FIRST CHAPTER!

THE MAGIC TWO TIMES TABLE

Let's pop into the Murderous Maths research lab, and meet our pure mathematicians. Today they are having a serious debate about which is the best times table.

Obviously there's a bit of a disagreement here, but we can always trust our mathematicians to discuss things in a sensible and responsible way.

Oh dear. In that case we'll just have to settle the argument for them.

The best times table is the two times table.

The reason is that if you can multiply and divide by 2, then you can multiply ANY two numbers!

Suppose you want to work out 37×94.

- Write the numbers at the top of two columns
- Keep dividing the number on the left by 2 and write the answers underneath. *Ignore any remainders.* Put * by any even numbers.
- Keep multiplying the number on the right by 2 and write the answers underneath.
- Cross out any lines with *
- Add up the remaining numbers on the right
- Ta-dah!

	37		94
*	~~18~~		~~188~~
	9		376
*	~~4~~		~~752~~
*	~~2~~		~~1504~~
	1		3008
	=		3478

THE MAD PEOPLE OF MATHS

If you just read the last little chapter you might think our murderous mathematicians are a bit mad. You'd be right. Here's one of them arriving for work all ready for another lovely day of doing sums. Notice the important details.

- RANDOM HAIR
- AN ORANGE NYLON SHIRT
- HALF A CHEESE AND CHUTNEY SANDWICH IN A POCKET THAT'S BEEN THROUGH THE WASH
- UNDONE SHOELACES

I ♥ M

Our mathematicians also do odd things such as laugh at bus timetables, or stare at the telly when it's off, but the best thing about them is that they are always happy. That's because the world will NEVER run out of numbers for them to play with, and what's more … numbers cost nothing!

Mathematicians have always been a bit strange. In fact, over the last 5,000 years they have included some of the maddest people in history! Are you ready for this? Here come some of the best…

Druid priests

Ancient people depended on being ready for summer and winter, so if you could keep count of the days and the sun's position, you could tell everybody what to do.

Nearly 5,000 years ago the old mathematicians worked all this out by building giant stone structures like Stonehenge.

They could even use their giant stones to predict eclipses! A *solar eclipse* is when the moon blocks out the sun in the middle of the day, and a *lunar eclipse* is when the Earth's shadow makes the moon disappear at night.

Everybody was terrified of eclipses, so it's hardly surprising that they thought the mathematicians must be powerful wizards. The mathematicians might even have made sacrifices to the sun and moon, which goes to prove ... mathematicians could get away with murder!

Weelll... You've got to make sacrifices in life AINTCHA!?

Can't I just have one sausage less for breakfast?

Thales

In ancient Greece they didn't have football or music videos, so instead one of the most popular things was ... maths! Lots of very clever people came up with new ideas, and around 2,600 years ago a man called Thales became a big star by showing that any angle in a semicircle is always a right angle. (That's the same as the corner of a square or 90°.)

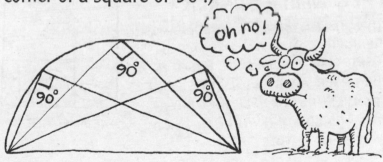

So was he a bit on the boring side? Not really. To celebrate, he went out and sacrificed a bull to the gods!

Pythagoras

Pythagoras carried on from Thales and formed a religious gang who worshipped numbers. They were called the Pythagorean Brotherhood and they had lots of rules, one of which was NEVER EAT BEANS.

OLYMPIC NEWS
STILL ONLY ONE GOLD COIN

PYTHAGORAS CALLS FOR TOTAL BEANS BAN SEE PAGE 2

PHEW WHAT A STINKER!

IT WAS REVEALED TODAY THAT A RANDOM BEANS TEST ON OLYMPIC CHAMPION FARTACLES HAS PROVED POSITIVE HE LATER ADMITTED EATING EIGHT PLATEFULS TO HELP HIM BREAK THE WORLD POLE VAULT RECORD WITHOUT A POLE.

WOW

WHAT A PONG

FARTACLES YESTERDAY

Pythagoras did lots of clever stuff including showing how harmonies in music work, but his biggest hit was proving that...

22

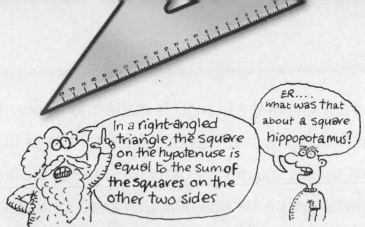

In a right-angled triangle, the square on the hypotenuse is equal to the sum of the squares on the other two sides

ER.... what was that about a square hippopotamus?

It's easier to understand what he meant by looking at a picture.

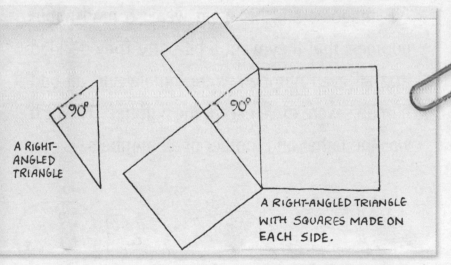

90°

A RIGHT-ANGLED TRIANGLE

90°

A RIGHT-ANGLED TRIANGLE WITH SQUARES MADE ON EACH SIDE.

A right-angled triangle is any triangle with a right angle, and the "hypotenuse" is the longest side, which is always opposite the right angle.

If each side of the triangle is made into a square, all Pythagoras said was that the total size of the two smaller squares is exactly the same as the big one. Sounds a bit dull, eh? Don't you believe it. This rule has helped people build bridges and skyscrapers. It even helps send rockets into space!

Pythagoras and his gang were so mad about numbers that they went a bit potty. They decided that all even numbers were female and all odd numbers were male except the number "1" which was the father and mother of all numbers...

…but when they were given problems that they couldn't solve with their nice neat numbers, they got so cross that they murdered people!

Maths wars

Hard to believe, isn't it? But yes, people did argue and even fight over who had the best bits of maths.

Because Pythagoras and his gang were so clever, there were others like the Eleatics who loved to upset them. The hero of the Eleatics was Zeno who used to invent *paradoxes*. These are things that look true but they can't be!

Zeno's paradox of the runner and the tortoise

A runner can move ten times as fast as a tortoise. However, if the tortoise starts one kilometre away, the runner will never catch him!

Think about it…

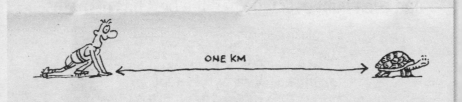

ONE KM

The runner runs one kilometre … but in the meantime the tortoise has moved one tenth of a kilometre, so the tortoise is still one tenth of a kilometre in front of the runner.

ONE TENTH OF A KM

The runner then runs the extra tenth of a kilometre ... but the tortoise has moved on just a little bit more.

ONE HARDLY ANYTHINGTH OF A KM

The runner runs the next little bit, but while he does so, the tortoise moves on just a tiny bit more ... and so on!

Even though the gap between them might be teeny weeny, *the runner will never quite catch the tortoise.*

GOING HOME

Of course, we know that the runner can catch the tortoise, but Pythagoras wouldn't have known how to work it out with his numbers. No wonder he got a bit upset!

Euclid

The great thing about brainy people arguing is that it makes them think harder and so their ideas get better and better. Around 300 BC another Greek called Euclid collected up all the clever stuff that the old maths gangs had discovered and put them in his book. It was called *Elements* and it's one of the most famous books ever written.

Euclid himself was a Pythagoras fan and he also came up with some theories of his own, including a rather natty fact about prime numbers. (A prime number won't divide by anything apart from itself and 1. The list starts 2, 3, 5, 7, 11, 13, 17, 19 … and Euclid proved that this list goes on for *ever*!)

Elements contained just about everything worth knowing about basic maths, and it inspired the next lot of mega mathematicians including…

The awesome Archimedes

Archimedes is one of the all-time greats. To realize how brilliant he was, you have to remember that all he used was a ruler and a compass. At the time there wasn't even a decent system of writing numbers down and doing calculations! Here are just some of the things he invented…

Giant lever systems

These were so powerful that his home town, Syracuse in Sicily, used them to grab and smash enemy ships! Archimedes  realized that lever power was so strong that he once claimed, "Give me a place to stand on and I can move the earth."

The Archimedean Screw

This is like a coiled pipe, and if you turn it round it seems to make water flow uphill! This invention is still used on farms and in factories today.

The Sand Reckoner

A mega number system. Back in those days there wasn't a good way of writing massive numbers so Archimedes invented one. It was based on the myriad, which is 10,000. He called a myriad of myriads (which is 100 million) "the first order of numbers". He then multiplied a myriad of myriads by itself a myriad myriad times (this is getting to be a very big number) and then multiplied that by itself a myriad myriad times. If we wanted to print the answer out on this page, it would start with a 1 followed by eighty thousand

.. a myriad myriads and one, a myriad myriads and two..

Honest! There's a bloke over there counting sand

NAAAAAH!

billion zeros ... and this page would be big enough for everybody in the world to stand on!

Archimedes said this number was "quite adequate".

Giant catapults

These were also used by the Syracuse army to smash invading forces.

A solar ray gun!

He is even supposed to have invented a set of mirrors that could focus the sun's rays onto enemy boats and set them on fire!

Brainstorm in a ... bath!

Despite all this brilliance, Archimedes is probably best known for jumping out of his bath and running down the road stark naked shouting, "Eureka!"

"Eureka" means "I have found it!", but what had he found?

Here's an experiment you can try yourself!

• Fill a bath right to the very top.

• Get into it very gently.

• Lie down very gradually so that you are just about floating.

• Guess what? The amount of water that has just sploshed on to the floor weighs the same as you do!

When you get into trouble, just tell people that you are testing Archimedes' *Principle of Hydrostatics*. This says that when you put something in water, the bigger it is, the more it gets pushed upwards. This leads to other tricks, such as you floating around in your overflowing bath.

Archimedes got the idea when the king asked him to check if his new crown was made from solid gold. The king suspected the goldsmith had mixed in some cheap silver, but the crown weighed the right amount so he couldn't be sure. Silver is lighter than gold, so if the crown was fake, it would be very slightly bigger. This would be extremely hard to measure, but Archimedes realized that the bigger crown would weigh less than a pure gold crown when it was underwater.

It turned out that when they put the king's crown under water, it DID weigh less that it should have done, which was very bad news for the cheating goldsmith!

Since he had that bath, Archimedes went on to discover a lot more about how and why things float or sink. If he'd skipped his bath and had a shower instead, we might never have known how to design ocean liners and submarines.

The ball in a tin

Archimedes's personal favourite invention wasn't his deadly catapults or his trick with the crown. To find out what it was, we'll go back to the island of Sicily in the year 212 BC.

The Romans are attacking the port of Syracuse where the 75-year-old Archimedes is hard at work…

HE'S ONLY THE GREATEST BRAIN ON THE PLANET! HE INVENTED THE ARCHIMEDES SCREW FOR RAISING WATER, LEVERS AND PULLEYS THAT LIFT MASSIVE OBJECTS, THE SAND RECKONER, WHICH IS THE BIGGEST COMPUTING SYSTEM EVER KNOWN, GIANT CATAPULTS, THE ARCHIMEDES PRINCIPLE... IN 2,226 YEARS HE'LL EVEN GET 15 PAGES TO HIMSELF IN A MURDEROUS MATHS BOOK!

39

Let's have a better look at that little sign he drew:

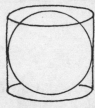

Archimedes proved that a sphere is exactly two-thirds as big as the smallest cylinder it will fit inside. In other words, if you have a solid ball that only just fits inside a tin can with the lid on, the ball takes up exactly $\frac{2}{3}$ of the space inside the can.

It might not sound as good as the sun-ray gun or the giant levers, but this simple fact is so important, for mathematicians it was like learning how to breathe!

A gruesome end to the Greeks

Although Archimedes lived in Sicily and was educated in Egypt, he was actually a Greek. After he died, the Romans took over the Greek empire and maths stopped being so trendy. Some people liked it, but they didn't get much encouragement.

One of the last was a very clever woman called Hypatia who used to attract huge audiences at her lectures around AD 400. Sadly the Christians thought she was a pagan and decided her fans needed discouraging. One day she was pulled off her chariot and dragged to the church where "her flesh was scraped from her bones with sharp oyster shells and her quivering limbs were delivered to the flames."

You never realized that being a maths teacher could be so dangerous, did you?

So... If you throw thirty Christians to seven lions how many would you have left?

I keep telling her. Hypatia, you're gonna upset the Christians

The maths mafia

One of the last ancient Greek mathematicians was Diophantus, and he has been nicknamed "The Father of Algebra". Algebra is a special way of setting out maths puzzles which have mystery numbers to work out. The mystery numbers are shown by letters, especially the letter x which gets used a lot. Some of these puzzles are dead simple but others are murderous.

Here are some algebra equations with their different names. Don't panic! You don't have to solve them unless you really want to.

Mm mmmmmm mm mmmm mmmm mmmm m mmmmm *

* NO THANKS I'D RATHER CHEW A SLUG

• Very simple algebra equation: $x = 6+2$

This is a LINEAR equation. Can you see that x equals eight? Dead easy!

• A bit harder: $2x^2+3x = 27$

The x^2 means this is a QUADRATIC equation.

• A lot harder: $5x^3+7x^2+2x = -16$

The x^3 means this one is CUBIC.

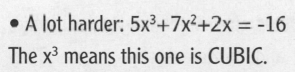

• Mercy! $3x^4- 5x^3+9x^2+2x = 43$

The x^4 means this one is QUARTIC.

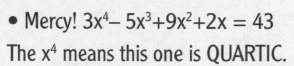

• Total brain destruction: $3x^5+41x^4-2x^3-x^2+7x = 3$

The x^5 means this is called, "ARGH, I think I'm going to be sick!"

AAAARGH!

More than a thousand years after Diophantus died, algebra started to really take off in Italy. Some pretty nasty characters including cut-throats and card cheats challenged each other with harder and harder algebra puzzles. They even used to bet money on who they thought would win – just like in a boxing match today.

ok champ - let him lead off with the linear equation then hit him with the quadratic - cubic combination

One match was between a man called Fior and another nicknamed "Tartaglia", which means "Stammerer". (It was hardly surprising, because when he was a boy someone shoved a sword through the top of his mouth!) They gave each other some very tough sums to do, and eventually Tartaglia won. He picked up the money, but he also revealed a new way of solving the really tough cubic equations.

Soon after this he was approached by Gerolama Cardano who must have been a really shady character. Among other things he was an astrologer, a doctor, an author, a gambler, a friend of the Pope and the father of a murderer. Cardano smooth-talked Tartaglia into giving away his secret method, and immediately ran off and published it in a book.

This same book included a way of solving the even harder quartic equations invented by Lodovico Ferrari.

At last ... a wimpish mathematician?

So far then, everybody involved in maths seems to have been pretty tough or got up to dodgy deeds or both. However, we've just come across Lodovico Ferrari ... the man who cracked quartic equations.

The amount of brain power needed just to understand quartic equations is massive, and solving them is twice as hard. Therefore you might think that Lodovico Ferrari was a timid little chap with a weedy moustache who always went shopping with his auntie.

Wrong! Lodovico used to drink, gamble, swear and fight, and in the end he was poisoned to death by his own sister.

Lodovico's last words

I thought that *!3⊙*3⊙ pasta tasted funny

Other oddballs

There are loads of other odd people who were famous mathematicians. Have you heard of the brilliant book *Alice in Wonderland*? It has the Mad Hatter's Tea Party and Tweedledum and Tweedledee, and the Queen of Hearts who played croquet with flamingoes and kept shouting, "Off with her head!". So what kind of crazy mind invented all this weird stuff? It was a mathematician called Charles Dodgson, although when he was an author he used the name Lewis Carroll.

How about the French teenager Évariste Galois?

Just before he died at the age of twenty, he scribbled down some algebra theories he'd thought of and years later people realized he was a maths superstar. However, he used to fail exams, fight teachers and was locked up for threatening the king. He died in 1832 in a duel over a woman ... so he was a *murdered* mathematician.

The list of strange characters in maths goes on and on (like the American professor who does all his thinking while riding about on underground trains!), but if we put all of them in here we wouldn't have room for anything else, so let's move on.

A tasty problem!

Our own pure mathematicians started getting strange ideas when they tried out Archimedes' principle with a football and a bowl of water...

49

How much of Archimedes' bathwater is in every glass of water you drink?

Water is made out of billions of tiny little things called molecules. A water molecule has two atoms of hydrogen stuck to one atom of oxygen. There are about 33,500,000,000,000,000,000,000,000 molecules in 1 litre of water.

First we need to know how many water molecules there were in Archimedes' bath. We don't know what sort of bath Archimedes had but we'll look into it.

YOU'RE NOT LOOKING IN **MY** BATHWATER!

Oh. In that case we'll assume it's like a modern bath, which holds about 200 litres. Therefore the number of water molecules in the bath is 200 × the molecules in 1 litre. That comes to 6,700,000,000,000,000,000,000,000 molecules.

We now have to assume that in the last two thousand years, these molecules have got into the sea and rivers, been evaporated by the sun, drifted along in clouds, come down in rain, got sucked up by plants and slurped up by animals, frozen into snowmen and generally mixed in evenly with all the other water on the planet.

Experts say that the total amount of water on Earth is about 1,260,000,000,000,000,000,000 litres. We want to know how many bathwater molecules there are in each litre of water, so we just divide the molecules by the litres:

$$\frac{6,700,000,000,000,000,000,000,000,000}{1,260,000,000,000,000,000,000,000} = 5,317,460$$

All drinks are mainly made of water, so for every litre you drink, you swallow over *five million* molecules of Archimedes' bath water!

Porky had visitors.

"It sure was nice of you to save our lives," said Weasel.

"Yeah, if you hadn't absorbed the blast, we'd never have made it down the corridor," said Half-Smile.

Porky had never been the hero before, and he felt good all over (except for his bottom which felt very sore).

A pair of high heels clicked across the tiles towards them.

"OK boys, break it up," said Dolly. "It's time to go to work."

"Says who?" replied Blade.

"Says the ten million dollars you owe," said Dolly. "With interest."

"I ain't interested in interest," said Blade.

The others all sniggered. Feeble as it was, it was one of Blade's better jokes.

"Yeah," said Jimmy. "If some mug wants to bail us for ten fat ones, that's his problem."

"Think it through, dummies," said Dolly. "You don't mess around with a guy who can pay out that much money. He'll have you straight back inside, and now your cell's blown up, the only place for you will be in Grimstate Jail. How would you like that?"

They wouldn't like it at all. Grimstate was so tough that the guard dogs were long-nosed, short-legged, African hunting hounds. (Ever since some prisoners mysteriously disappeared, the people in charge stopped using the word "crocodile".)

"Every meal is leftovers from the meal before!" wailed Porky. "There's one potato there that has been dished up three times a day for the last 19 years."

"So how many times is that?" asked Dolly.

"How should I know?" asked Porky.

"It's just simple maths!" said Dolly. "Three times 365 times 19."

"Two-oh-eight-oh-five," snapped Numbers.

"Maths is for cissies," said Blade. "Who needs it?"

"The interest on this ten million is 15 per cent every week," said Dolly. "So every day you do nothing, you owe more money."

"Every day?" gasped Blade. "How much more?"

"If you want to know, be a cissy and work it out," said Dolly.

The mob all put their heads together and started to mutter.

Dolly couldn't bear it any more.

"Roll over, big guy," she ordered, then pulled out her lipstick.

"What for?" moaned Porky.

"I need a drawing-board."

The big man twisted round in his bed to reveal the huge white bandages on his rear end.

"Listen up," said Dolly. "Fifteen per cent interest a week means that for every hundred dollars you owe, after one week you have to pay an extra $15.

"Is that all?" sneered Blade. "A lousy 15 bucks a week?"

Dolly wrote the figures out in lipstick on the bandages.

"That's if you owed a hundred dollars," said Dolly. "But you owe ten million. How many lots of a hundred is that?"

"One hundred thousand," snapped Numbers. "One oh-oh-oh-oh-oh."

"And we have to pay $15 interest on every one of those?" asked the Weasel.

"That's right," said Dolly. "So for every week…"

"One million, five hundred thousand dollars of interest," interrupted Numbers.

"OK, you made your point," said Blade. "We start work tomorrow."

"Are you sure you can afford to wait?" said Dolly. "That works out as over $200,000 interest every day. This time tomorrow you'll be nearly a quarter of a million poorer."

"We start today," said Chainsaw. "I might not know much about maths, but I know about being broke and that's what we're gonna be if we don't get moving!"

"Good," said Dolly. "Get down to the freight yard. I'll see you in the signal box in one hour."

To be continued...

DEADLY DOMS

So there you are popping back from the shops with a pizza when…

"Aha!" comes an evil voice. "Gotcha!"

Suddenly you find yourself swinging high above the ground. It turns out that your arch-enemy Professor Fiendish has just driven by in his crane and grabbed you with the hook.

"Come with me," he sneers. "I've got a puzzle that you'll never solve. Har har!"

Soon you find yourself in a cell which has a skeleton chained to the wall. In front of the skeleton is a chessboard and a box of dominoes.

"The chessboard has got eight squares along each side," says the professor. "So how many squares does it have altogether?"

Quickly your super brain works out the answer. It's 8×8 = 64.

"Sixty-four!" you say sounding bored. "Honestly professor, is that your puzzle? How utterly feeble."

"That was NOT the puzzle!" says the professor crossly. "This box has exactly enough dominoes to cover the chessboard, and each domino covers two squares."

"How jolly interesting," you say.

"So how many dominoes are in the box?"

Once again your super brain solves the problem in a flash. It's just $64 \div 2 = 32$.

"Thirty-two dominoes," you say with a casual yawn. "Can I go now? My pizza is getting cold."

"No! That was NOT the puzzle either," snaps the professor. "I'm going to chop two squares off the chessboard, and take away one of the dominoes."

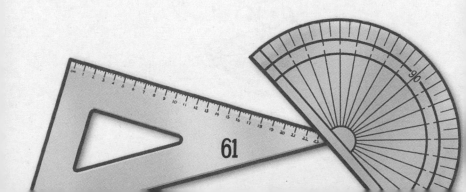

The professor takes a knife and chops away two squares from opposite corners of the board. The squares are both white.

"There are now 62 squares left and 31 dominoes," says the professor. "Can you cover all the squares now?"

Oh dear, this is where it gets nasty! Obviously the owner of the skeleton couldn't solve the puzzle, but can you?

Different sorts of 'ominoes

Forget about the spots on dominoes, just think about the shape. A domino shape is made up of two little squares joined together. There, that was a quick think, wasn't it?

Suppose you only used *one* square, it wouldn't really be a domino would it? In fact you could call it a *monomino* because *mono* usually means "one".

How about using *three* squares? You get a *tromino*, but something slightly exciting happens here. There are two different sorts of tromino — you can have a straight tromino where the squares are all in a line, or you can have a bent tromino, where the squares make a little corner.

With *four* squares you can make *tetrominoes*, and there are five different types carefully drawn out here.

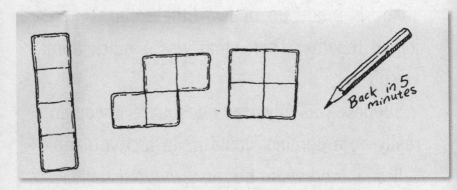

Oh dear. The artist has just popped out for a custard sandwich, so you'll have to draw the last two tetrominoes in yourself. You'll know the five tetronimo shapes if you've ever played the game Tetris.

Probably best of all, with *five* squares you can make *pentominoes*.

There are 12 different pentominoes and here you can see them arranged into a neat 6×10 rectangle. It looks simple, doesn't it?

There are lots of games to play with pentominoes. You can download pentominoes as an app, but it's much nicer to make your own set of real shapes to play with. The easiest way is to copy this page on to some card (and make it bigger if you can) then cut out the shapes and colour them in.

Here are some wicked challenges for you:

1 Arrange the 12 pentominoes back into a 6×10 shape like before. There are 2,339 different ways of doing this, but if you can find more than three or four then you're doing well!

2 Arrange the 12 pentominoes into a 5×12 shape. There are 1,010 ways of doing this, but it's quite tricky!

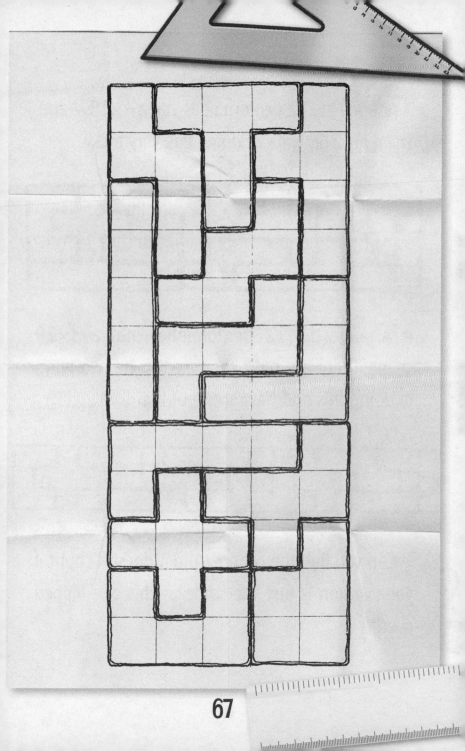

3 Arrange the 12 pentominoes into a 4×15 shape. There are 368 ways of doing this. Very tricky.

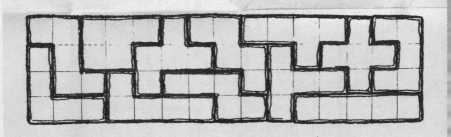

4 Arrange the 12 pentominoes into a 3×20 shape. There are only supposed to be two ways of doing this and here's one of them.

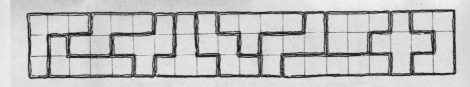

Can you find the other? (And it doesn't count if your version is just the same as this one flipped over!)

5 Pick any one of the shapes. Using nine of the other shapes, can you can make a big version of the shape you have picked…

6 Finally, here's a great game to play with a friend. You need your pentominoes and a square board (like a chessboard) marked out with 64 squares so that a pentomino covers five squares.

• The first player picks a pentomino from the set and puts it on the board covering any five squares.

• The other player picks another pentomino and puts it on the board, but it mustn't overlap the first pentomino.

• Keep taking turns until somebody can't go.

• Whoever puts down the last pentomino wins.

A rule to break

You're not supposed to turn the pentominoes over in any of these puzzles (i.e. you're not supposed to make the shape on the left into the one on the right)...

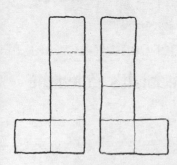

...but we think that's a silly rule, so just ignore it. You might even find more solutions to the puzzles!

Enough about pentominoes.

Hexominoes use *six* squares and there are 35 different shapes. Here are just a few of them...

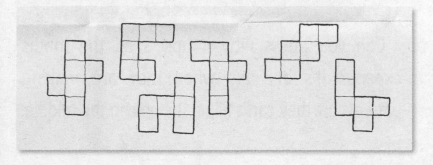

Heptominoes use *seven* squares and there are 108 different shapes including these…

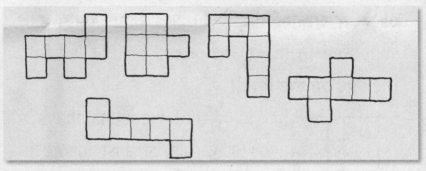

One heptomino is rather special. It's called the "harbour" heptomino…

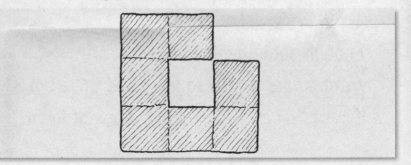

Can you guess why it upsets all the maths experts? It's because when they are making patterns up, they can't fill up the hole in the middle.

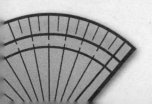

Maths experts are funny people. If you give them an equation like Endean's Determination of the Invariance of the Hubble Constant:

$$T_o = \frac{2L\left[S_o\left(T_o\right) - \sin S_o\left(T_o\right)\right]^{\circ}}{\sin^2 S_o\left(T_o\right)}$$

AHH!

...they twinkle like snowflakes. But show them a little drawing of seven squares in a circle and they lock themselves in the bathroom and chew their toes.

ARGHH! TAKE IT AWAY! MUMMY!

Whoops! We mustn't forget that we haven't solved Professor Fiendish's chopped chessboard puzzle yet. We'd better do it quickly before your pizza turns to rubber.

The answer is that the puzzle is impossible, but the professor will have to let you go if you tell him why. The answer is rather neat!

When you put the dominoes on the chessboard, each domino covers two squares next to each other. This means each domino has to cover one black square and one white square. After the professor chopped off two white squares, the board was left with 30 white squares and 32 black squares.

When you've put 30 dominoes on the board, you will have covered up 30 black and 30 white squares. You will always leave two black squares uncovered. It doesn't matter how you arrange the dominoes, and it doesn't matter where the two uncovered squares are, *they both have to be black!*

Obviously there's no way you can cover two black squares with one domino (you're not allowed to break it in half), so the puzzle is impossible!

Quick! Before the professor gives us another diabolical problem, let's dash off to the next chapter...

THANKS TO YOU IT'S GONE SO STALE I'D RATHER EAT THE BOX!

THE IMPOSSIBLE RACE

Here are some speeds:

• The fastest human sprinters can do about 10 metres per second (10 m/s).

• Light travels at 300,000 kilometres per second (300,000 km/s).

• The fastest land animal is the cheetah, which can run at about 100 kilometres per hour (100 km/h).

• A really turbo-charged snail might just reach 50 metres per hour (50 m/h)

• New York is moving away from London at about 20 millimetres per year, thanks to what's called continental drift. (No wonder air fares keep going up with planes having to fly further.)

So if they all had a race, who'd win?

It's all a murderous mix-up, isn't it? Are kilometres per second faster than metres per hour? Do millimetres per year beat metres per second? In other words, which speed is the fastest?

SORRY, BUT WE NOW INTERRUPT THIS BOOK TO BRING YOU AN URGENT NEWSFLASH!

REPORTS HAVE JUST COME IN OF AN **ALIEN INVASION FLEET!** THE EVIL GOLLARKS FROM PLANET ZOG HAVE LANDED AND THEIR BATTLE CRUISER IS PREPARING TO ADVANCE!

PREPARE FOR THE **UTTER DESTRUCTION** OF **ALL KNOWN LIFE ON EARTH!**

SCLOINK

HOWEVER, WHILE YOU'RE WAITING FOR YOUR GHASTLY FATE AT THE HANDS OF THE GOLLARK AVENGER DROIDS, WE SUGGEST YOU RELAX AND ENJOY SOME MORE **MURDEROUS MATHS...**

Speed is pretty simple. All you need to know is this:

SPEED is the DISTANCE you move divided by the TIME it takes. If you want to be lazy you can just put:

$$\text{Speed} = \frac{\text{Distance}}{\text{Time}}, \text{ or even } S = \frac{D}{T}$$

When you see a speed like "ten kilometres per hour" it just says that if you want to go ten kilometres, then it will take you one hour.

OK, now get ready for a deep question…

What is the meaning of per?

IT MEANS I JUST **LOVE** HAVING MY TUMMY TICKLED!

NO, NOT **PURR!** WE'RE TALKING ABOUT **PER!**

Per means *for each*, so if you move at 30 km *per* hour that means you move 30 kilometres for each hour you travel. When it comes to sums, *per* means *divided by* and so you can swap it for a little dividing line like this "/". So 30 km per hour gets written as 30 km/h.

Suppose you could skate along at 28 kilometres per *two* hours? There's nothing actually wrong with it, it's just that you don't expect any numbers after the "per". Remember that speed is distance divided by time, so you can turn "28 km per two hours" into a sum:

$$\text{Skating speed} = \frac{28 \text{ km}}{2 \text{ hours}}$$

Now you work out 28÷2 and get:

$$\text{Skating speed} = \frac{14 \text{ km}}{1 \text{ hour}} \dots \text{ or } 14 \text{ km/h}$$

80

Instead of saying "14 kilometres per hour", some people just say "14 kilometres *an* hour". Don't be confused, they mean the same thing.

WE'LL GIVE YOU FURTHER INFORMATION WHEN WE HAVE IT. IN THE MEANTIME THE PUBLIC ARE ASKED NOT TO PANIC, SO RELAX, SMILE, THINK SWEET THOUGHTS AND LET'S HAVE SOME MORE MURDEROUS MATHS...

Let's go back to our race with everything going at different sorts of speeds. We've got millimetres, seconds, hours, kilometres ... so how do we see which one is fastest? The secret is to choose which sort of speed we like the best and then change all the others to fit it.

"Metres per second" looks nice and simple so we'll use that. We can write it as m/s.

We're told that the sprinter can reach 10 m/s. Hooray! We don't need to change that one.

Light travels at 300,000 kilometres per second, so we need to turn these kilometres into metres. As one kilometre = 1,000 metres, this means that 300,000 kilometres is the same as 300,000×1,000 = 3,000,000,000 metres. So now we know that light travels at 300,000,000 m/s.

The cheetah runs at about 100 kilometres per hour. Once again we multiply by 1,000 to change the kilometres into metres. We find that cheetahs run at 100,000 metres per hour.

Hang on … *per hour* is no good! We need to know how many metres a cheetah can run *per second*. To work it out, we'll write this speed like a sum. (Here we go again! Remember speed equals distance divided by time.)

$$\text{Cheetah speed} = \frac{100{,}000 \text{ metres}}{1 \text{ hour}}$$

Now we need to know how many seconds there are in an hour, but that's not too hard. There are 60 minutes in an hour, and each minute has 60 seconds, so in one hour there are 60×60 seconds which comes to 3,600. Now we can swap the hour in our sum for 3,600 seconds:

$$\text{Cheetah speed} = \frac{100{,}000 \text{ metres}}{3{,}600 \text{ seconds}}$$

If you work this out on a calculator, make sure you put in exactly the right number of noughts!

Cheetah speed = 100,000÷3,600

= 27·77777777777777777 m/s

What a boring lot of 7s! But if you read *The Secrets of Sums* you'll see that we can round this off and say the cheetah runs at 27·78 m/s.

 The snail can reach 50 metres per hour, so will the snail work out to be faster than the cheetah? The only way we can be sure is to change its speed to m/s, so off we go!

The distance is already in metres so we don't need to change that, but we do need to convert the hour to seconds. We do what we did for the cheetah and divide by 3,600:

$$\text{Snail speed} = \frac{50 \text{ metres}}{3,600 \text{ seconds}}$$

$$= 50 \div 3,600 = 0.0138888888888 \text{ m/s}$$

The eights look pretty dull too, so we'll round them off to find our snail goes at a rather nifty 0.01389 m/s.

Finally, New York moves at 20 mm per year, so we'll write out it like this:

$$\text{New York speed} = \frac{20 \text{ mm}}{1 \text{ year}}$$

First we have to change the 20 little millimetres to metres. For this we divide by 1,000 so 20÷1,000 = 0·02 metres. Then we have to convert years to seconds. How many seconds are there in a year, eh?

We already worked out there are 3,600 seconds in 1 hour. A day has 24 hours so there are 3,600×24 = 86,400 seconds in 1 day. There 365 days in a year, so that makes 365×86,400 = 31,536,000 seconds in a year.

Of course, if you want to be a real pain, you now say this:

WHAT ABOUT A LEAP YEAR? IT HAS 366 DAYS...

This is the sort of remark that makes maths teachers wish they'd gone to join a circus instead.

Unless you're a weirdo who counts how many cornflakes you ate for breakfast, then for sums like this, one extra day in a whole year doesn't matter.

Right then, back to New York. We divide the metres by the seconds to get this sum:

$$\text{New York speed} = \frac{0.02 \text{ metres}}{31,536,000 \text{ seconds}}$$

And when you divide that out you get: New York moves at 0·0000000006342 m/s.

At last! All our speeds are now in metres per second, so let's see how they compare.

NEW YORK:
0.0000000006342
M/S

CHEETAH:
27.78
M/S

LIGHT:
300,000,000
M/S

SNAIL:
0.01389
M/S

SPRINTER:
10 M/S

PANT PANT PANT PANT

1

Well surprise, surprise, light travels a lot faster than New York. Worth all the effort to find that out, wasn't it?

ONCE AGAIN WE INTERRUPT TO BRING YOU SOME IMPORTANT GOLLARK DATA: ONE GLOMP IS THE SAME AS 19 EARTH METRES. ONE MNULT IS THE SAME AS 3 EARTH DAYS

1 GLOMP = 19 METRES

1 MNULT = 3 DAYS

AS THE CRAFT CONTINUES TO ADVANCE WE CAN ONLY WAIT AND WONDER — WHAT ARE THE EARTH'S CHANCES OF SURVIVAL?..

RUMBBBBLE

RUMBBBLE

THE END IS NIGH

By now you'll be getting pretty scared about the Gollark invasion, so let's see how long we have to live. We know the Gollarks' speed is 180 glomps/mnult and we've got all the information we need to change this into metres per second.

180 glomps = 180×19 = 3,420 metres

1 mnult = 3×24×60×60 = 259,200 seconds

$$\text{Gollark speed} = \frac{3,420 \text{ metres}}{259,200 \text{ seconds}}$$

When we work out 3,420÷259,200 we find that the speed of the Gollark battle cruiser is 0·01319 m/s.

Hey! Just a minute, we've already found that snails move at 0.01389 m/s.

Good grief! Snails are slightly faster than the Gollark battle cruiser.

WE INTERRUPT YOU ONCE AGAIN TO BRING YOU INCREDIBLE DEVELOPMENTS IN THE GOLLARK SAGA...

IT SEEMS THAT NOT ONLY CAN SNAILS CATCH THE BATTLE-CRUISER...

EEEEk!

SLURP

GUZZLE

MMM!

SCOFF

OOOM!

CHOMP

» GULP

SCRUMP

IT'S CRISPY!

..IT ALSO TURNS OUT THAT THE CRUISER TASTES LIKE LETTUCE!

Earth is safe once again but it could have been a very different story – if only the Gollarks had known about murderous maths!

ONE-SIDED PAPER

Dear Murderous Maths,
   I have just read the chapter heading in your book which says "One-sided paper".
   One-sided paper? Impossible! In all my years of teaching maths I've never heard of such rubbish.
   If you were in my class I would make you write out 10,000 times in very big letters "I must not waste paper".

   Yours unimpressedly,

   P.J. Blenkinsop

Sorry P.J. Blenkinsop, but you aren't going to like this much.

Let's think about a normal piece of paper. (Ooh yes please, that sounds really interesting.) It has two sides, hasn't it? And round the edge of the page is ... the edge! This is clever stuff, isn't it?

Now let's suppose that Professor Fiendish has invented a new *killer colour*...

HEH! HEH! ONE GLANCE AT THIS SEPTIC SHADE IS ENOUGH TO MAKE YOUR NOSE EXPLODE... AND I'VE DYED YOUR PIECE OF PAPER RIGHT THROUGH WITH IT!

KILLER COLOUR

You have to quickly paint the whole piece of paper black so that none of the killer colour shows at all. This is what you'd need to do:

1 Start on one side of the paper and paint across to the edge.

2 You then have to go over the edge to paint the other side. The reason you have to go over the edge is because the paper has two sides. However, if the piece of paper only had one side, you could black out the killer colour completely without going over an edge!

How to make a piece of one-sided paper

In the 1960s (when nobody had colour telly and teachers had little flowers painted round their belly buttons) this sort of thing was called *modern maths*. Now that we've all got colour telly and the teachers just have fluffy belly buttons, it's probably called *ancient maths*, but it's still a fun thing to do.

What you need are two long strips of paper, and some glue or tape.

1 Make one strip of paper into a big circle by gluing the ends together. (As if you were making a big bracelet.)

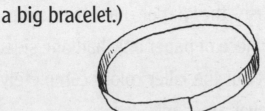

2 Do exactly the same with the other one, but before you stick the ends together, twist one end round so that it is upside down.

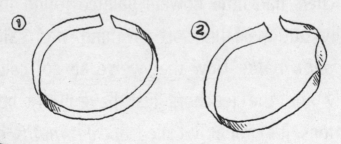

This second circle is actually a one-sided piece of paper! It's called a Möbius strip, after the bloke who invented it.

1 Take your first circle, and draw a line along the middle of the strip until you get back to where you started. You'll find that one side of the paper has a line on, and the other hasn't.

2 Now draw a line along your second circle and *keep going* until you get back to where you started. You'll find you've drawn your line on what was both sides of the paper. That's because you've joined the two sides up, so this paper only has one side! If this paper had been in "killer colour" you could have blacked it all out without going over the edge.

There's something else funny about this. One of your circles has two edges, but the other circle only has one! To test this you need to find two ants who are madly in love.

Get your first paper circle and put one ant on each edge. Tell the ants that they can go and see each other, but have to walk along the edge to get there. What a mean trick – they'll never do it!

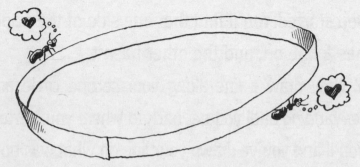

However, if you put your ants on each edge of the second circle, they can meet up by just walking along. This is because the one-sided piece of paper only has one edge!

A magic trick

If you haven't got wobbly hands, get some scissors and try this. Take the first circle and cut along the line that you drew. (Be careful of the ants because they might have cheated and be kissing in the middle.)

You'll end up chopping your circle into two thinner circles, but that was obvious wasn't it? If you chop *anything* down the middle, you get two bits, don't you?

Here's the weird bit! Cut along the line you drew on your second circle. Even when you've chopped the whole thing into two – what do you get?

Here's another fun thing to do with one-sided paper.

Make another circle with a twist in it, then draw a line round it that's quite near the edge. You'll need to go right round the circle twice until it joins back on itself.

Can you guess what you'll get when you cut along the line?

One final freaky illusion

Here's a Möbius strip trick to really amaze your friends – and even yourself!

1 Get three long strips of paper. One of them should be slightly wider than the other two and a different colour.

2 Put them together with the wide one in the middle like a long sandwich. You'll find life is much easier later on if you use a few very small bits of Blu-tak to hold them together.

3 Put the ends of the sandwich together and add a twist, as if you were making a three-layered Möbius strip.

4 Stick the ends of the wider piece of paper together, and then stick the other loose ends together. You should have three separate joins when you are finished.

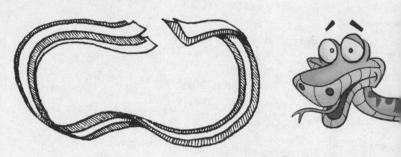

5 Show what you have made to a friend. Point out that there are two loops with a different coloured loop in the middle separating them. As the coloured loop is wider, you can see it sticking out all the way round, so there is NO WAY the two plain loops are even touching each other!

6 Ask your friend – without tearing the paper – to pull the two plain loops away!

Dear Sir,
   I have been trying your silly experiments with one-sided paper.
   I found two ants in love, and then spent a year training them to walk along the edge of the paper. Unfortunately, it turns out that ants only live for a few weeks and so I have wasted many months talking to two dead ants, and trying to lead them round in circles with drops of honey on the end of a matchstick.
   I am sure many more of your readers have done exactly the same as me because of your irresponsible book. You should be ashamed of yourself.
   Yours crossly,

   PJ Blenkinsop

Dolly was already waiting at the top of the stairs to the signal box as the six men arrived.

"Where's the big one?" she asked as they made their way up.

"He stayed back for lunch," replied Chainsaw.

"And when Porky stays for lunch, there's no shifting him."

"I guess he's done his bit," admitted Dolly. "You guys get inside. There's someone you need to meet."

The signal box smelled of old oil and wood rot. A large display board covered one wall, while on the other side a massive window overlooked a maze of railway tracks.

"Gee, look at all the big levers," said Half-Smile Gabrianni.

"They work the points and signals and stuff," grinned a small woman in a large boiler suit. "Pull one wrong and this whole place will pile up into a junk yard."

"Guys, meet Harri," said Dolly.

"Harry?" they all said.

"Harri as in Harriet," drawled the small woman. "Anyone want a chew of my spitting tobaccy?"

"No thank you kindly, ma'am," they muttered as they shuffled backwards into the far corner.

"What's with all the pretty lights?" asked One-Finger Jimmy, pointing at the wall display with his one finger.

"This chart shows the rail track from here to Fort Knocks," said Dolly. "And the stops in between."

"And the lights show how the points are facing," said Harri. "As worked by me and my levers."

"OK, now listen up," said Dolly. "The Knocks Express leaves this yard tomorrow morning at 5.00 a.m. It travels at 30 miles per hour and reaches the Fort at 6.40 a.m. precisely. It'll be carrying 15 million bucks."

"Not to mention about 200 guards," said Blade.

"Not this one," said Dolly.

"No guards?" asked Chainsaw. "But what's to stop anybody just leaping aboard and grabbing a stash of cash?"

"This 15 million is in coins," said Dolly.

"But that would take days to unload!" said Blade. "Supposing we stopped the train, how would we get it all away before the police arrive?"

"Tell 'em, Harri," said Dolly.

"Five miles before Knocks, there's a set of points where an old mining track comes off the main line," said Harri. "Those points still work, and I'm the only person who knows it."

"So all you guys have to do is take over the Express, run it down the mining track and hide it," said Dolly.

"Yeah, and once the points are set back, I'll cut the cable," said Harri. "Nobody will ever find the train, so you can empty it in your own good time."

The men glanced at each other. It seemed too easy.

"OK, but how do we stop the train?" asked Jimmy. "I guess they don't pull up for hitchhikers."

"Yeah," agreed Weasel. "And the old broken-down-car-stuck-across-the-line trick is so obvious they'll just smash right through it."

"Cows," said Dolly.

"Cows?" they asked.

"There's a cow field right by the points," said Dolly. "Just before the train arrives you open the gate and herd them out across the line."

"But suppose the beefs get bored waiting and just wander off?" asked Blade. "Hell, none of us is cowboys, we can't keep them in place!"

"You'll just have to get your timing right, won't you?" said Dolly. "You'll have to work out exactly when the train will be coming."

"Oh no," said Blade. "Not *more* maths!"

To be continued...

So how long will Pongo have to wait until he can use the same calendar again?

Most years have 365 days in them, and there are 7 days in a week. To find out how many weeks there are in a year, it's 365÷7 which gives 52 weeks with one extra day left over. If January 1st is on a Monday this year, the extra day will move it along to a Tuesday next year.

That's why there are seven different versions of the normal calendar, and each of them starts on a different day of the week.

What makes it even more complicated is that February usually has 28 days, but every fourth year there is an extra day – Feb 29th. These years are called leap years because if Jan 1st is a Monday then the next Jan 1st will leap over a day and be on

a Wednesday. There are also seven different leap year calendars which start on different days of the week, so in total there are 14 different calendars.

Every 28 years we can use each of the seven normal calendars three times, but we only use the seven leap year calendars one time each.

The strangeness of Friday 13th

Some people suffer from (take a deep breath…) *paraskavedekatriaphobia* which is a fear of Friday

13th. Sailors used to think it was unlucky to set sail on a Friday, and lots of people think 13 is an unlucky number, so if a Friday is on the 13th, that's *two* lots of bad luck together! It's also the day that witches are supposed to meet up.

So how often do we get a Friday 13th?

The way calendars work out, there has to be one Friday 13th every year and never more than three. What's more, because the days in a year never divide by 7, the weekdays should never fall into a pattern. Therefore the 13th should have an equal chance of being *any* day of the week...

...but Friday 13th is slightly spookier than that!

OOH!

This is where the maths get murderous!

A year is how long it takes the Earth to go round the sun. It's almost exactly 365·25 days. The extra 0·25 (or $\frac{1}{4}$) of a day adds up and that's why we have an extra day every four years. However the *exact* number of days is 365·242199 which is very slightly less.

To make up for this, if the year ends in 00, although it *should* be a leap year, it isn't unless the year divides by 400! Therefore 2100, 2200, 2300 will not be leap years, but 2400 will be. So don't forget that when we get to 2400, will you?

This means that the total number of days in 400 years (which includes 97 Feb 29ths) is 400×365+97 = 146,097. This number divides exactly by 7 which means that the weekdays *do* fall into set pattern. In 400 years there are 4,800 months and it turns out that:

684 months have the 13th day on a Saturday

687 months have the 13th day on a Sunday

685 months have the 13th day on a Monday

685 months have the 13th day on a Tuesday

687 months have the 13th day on a Wednesday

684 months have the 13th day on a Thursday...

... and 688 months have the 13th day on a Friday!

Therefore the 13th is very slightly more likely to be on a Friday than any other day of the week!

If you got through that murderous bit of maths, you deserve a little treat, so here's something cute... Which months have 31 days?

Close your fists and put your hands together.

Imagine the months written out on your knuckles and the gaps between them. The months that land on the knuckles have 31 days!

MONTHS WITH 31 DAYS SIT ON TOP OF YOUR KNUCKLES

How many people in the world share your birthday?

We can't work this out exactly because the number of people in the world keeps growing, but we can get a rough idea. Let's say the total population is about 7,200,000,000 and to make the sums easier we'll say there are about 360 days in a year. This means the number of people who have their birthday on each day is roughly 7,200,000,000÷360 = 20,000,000. This means that most of us share our birthday with about *twenty million* other people!

If you were born on Feb 29th, then you only have one proper birthday in every four years which is 1,461 days, and it works out that you only have to share your birthday with about *five million* people.

Maybe you should throw a party and meet them all!

116

LITTLE NUMBERS WITH SUPERPOWERS!

Here's a nice little sum: $13+5 = 18$

This one isn't too bad either: $13×5 = 65$

But what do you think this one makes: $13^5 = ?$

That little 5 might look a bit feeble, but 13^5 is called *13 to the POWER of 5* and it means five 13s all being multiplied together. You get $13×13×13×13×13$ which comes to 371,293!

Any little number up in the corner by another number is called a power, and it can give you some murderously massive answers!

How to scare a scientist

Powers can be pretty scary, especially when it comes to things like *bacteria*.

Bacteria look like alien maggots and they can live almost anywhere. If you'd like to see some, turn back a few pages and look at Pongo's burger bar. It's covered in them! You'll have to look very carefully though because bacteria are so tiny you could fit millions on the head of a pin. You've also got some friendly bacteria crawling round inside your guts. Can you feel them?

There are thousands of different sorts of bacteria. Most of them are harmless, but if you get enough of the dangerous ones, they can be lethal! Scientists are always trying to invent new drugs to fight dangerous bacteria, but they have two *big* problems...

• Although a drug might kill billions of bacteria, occasionally one or two become mutants. This means they can resist the drug and survive.

• Even if the mutant bacteria are lethal, one or two can't hurt anyone on their own. The trouble is that they can reproduce themselves rather quickly … thanks to powers!

The bacteria guide to reproduction

1 Grow longer and split in half.

2 Both halves grow longer and split in half again.

3 Do this *every 10 minutes* (or even faster).

If you start with just one lonely little bacterium:

• in 10 minutes you will have 2

• in 20 minutes you will have 2×2 (or 2^2)

- in 30 minutes you will have 2×2×2 (or 2^3)
- in 60 minutes you will have 2×2×2×2×2×2 (or 2^6)
- in 24 hours (which is one whole day) you will have 2^{144} bacteria!

If you work out 2 to the power of 144 you'll find it comes to about: 22,300,000,000,000,000,000, 000,000,000,000,000,000,000,000. That's how many bacteria you can get in just *one day*. Murderous!

Luckily, it's only the fastest types of bacteria that can grow and split in 10 minutes or less. Most take around half an hour but even so, in one day you would get 2^{48} of them which is about 281,000, 000,000,000. In two days you would get

$281,000,000,000,000^2$ which comes to about 79,200,000,000,000,000,000,000,000,000.

If this was a lethal bacteria…

- and there was nothing to stop it reproducing
- and it managed to spread itself
- and there was nothing to destroy it

…this would be enough to kill everybody in the world.

No wonder scientists are worried!

By the way, if you think it's a bit freaky fitting millions of bacteria on the head of a pin, the next chapter is even stranger!

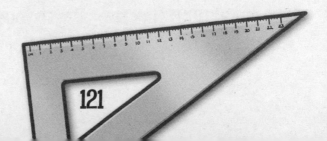

WHAT IS THE LONGEST LINE
YOU CAN DRAW ON THE HEAD OF A PIN?

Pinheads are only tiny, so you might think your line could only be about 1 mm long. Have a look.

Isn't that a neat little picture? However we do realize that there are one or two *Murderous Maths* readers who might not be able to make out all the tiny details, so we've made it bigger for you.

If the line has to be straight, then 1 mm is the answer. But what if the line doesn't have to be straight? Let's see what a zigzag line looks like.

THIS LINE IS ABOUT 12 mm.

As you can see, if we try to fill the space up then the line will be longer. And if the line is thinner – it could be a LOT longer!

THIS LINE IS ABOUT 70 mm.

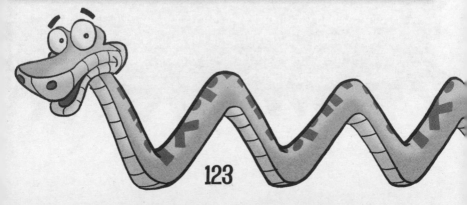

It's no good, we need to make this even bigger.

VERY IMPORTANT THING TO DO
Before reading any more, you MUST find the two ends of the line in the picture. Remember, one end is not good enough. Even our most murderous artist cannot draw a one-ended line, so make sure you find them both.

There are two strange things going on here.

1 If you make the line thinner, then you can make it longer and still have room to fit it on the pinhead.

2 There is NO LIMIT to how thin a line can be.

Here's the really scary bit. You could draw a line long enough to reach the moon and back, but if it was thin enough it would still fit on the head of a pin! So the longest line you can draw on the head of a pin is ... *as long as you like!*

This is because a real line only has length. It has absolutely no thickness at all. The fancy way of describing this is to say that a line is "one dimensional".

WHO ARE YOU CALLING ONE DIMENSIONAL?

ARE YOU SAYING WE'RE SHALLOW?

YOU'RE ASKING FOR A GOOD KICKING, POSKITT!

The confusing thing is that when we mark lines on paper we have to give them a little bit of thickness or we couldn't see them. The thickness might only be tiny, but it means that we haven't drawn a real line, what we've drawn is a coloured-in *area*. Although a real line would not take up any space on the page, areas always do.

How to put a real line on this page

Would you like to put a line on this page that only has length but no width at all? It might seem impossible but the trick is that you don't use a

DON'T YOU DARE CUT THIS BOOK OR I SHALL LOSE MY TEMPER!

pencil or a pen. You use scissors.

He's right, you shouldn't cut books, but just suppose you were to do one very neat cut from the edge of the page between these two drawn lines.

_____

_____

When you hold the page down flat, you'll just be able to make out the very thin line where the cut is — but it has no thickness! It will be a perfect one-dimensional line.

Have you done a cut? Quickly, put the scissors away and let's hope nobody gets hurt. Now turn the page over.

Oh dear! When you saw this book was a *Murderous Maths* book, you never realized how murderous it could be did it? Well, let that be an important lesson to you, never ever cut pages in books.

Quick! Change your clothes, put on a false beard, grab a fake passport and hurry away to the safety of the next chapter…

How to climb through a postcard

Although it's almost impossible to draw a long line on the head of a pin, there is another trick like it that you can try.

Show a friend a postcard or a birthday card and say you can cut a hole in it big enough to climb through. Your friend will think you're nuts, but here's how you cut the hole…

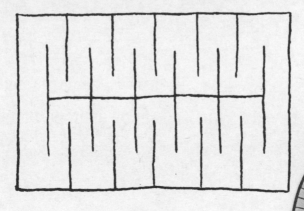

The card will open up into a long jagged circle which you should be able to step through. The strange bit is that if you really take your time and cut more and more slots in the card, you can make the hole big as big as you like!

THAT'S CLEVER... BUT WHAT'S IT GOT TO DO WITH MATHS?

WHO CARES? AT LEAST IT HASN'T GOT ANY SUMS IN IT!

Do you know a calculator clot? That's a person who can't do anything without a calculator. If you want to test somebody, ask them this question:

"If I have six calculators and lose two ot them, how many will I have left?"

If it's a clot they'll say, "Ooh I dunno, where's my calculator?"

The silly thing is that sometimes calculators are completely useless at sums.

The cake, the clot and the calculator

Suppose you want to share a cake with six friends.

How much cake does each person get? You need to divide one cake by seven people (don't forget to include yourself). At this point the calculator clot will pull out a calculator and put in $1 \div 7$. He will then say that each person should get 0·1428571 of the cake.

Can you imagine going up to a cake and trying to cut a piece which is 0·1428571 big? Of course not.

The best thing to do is tell the clot he should eat his calculator instead. Now you can divide the cake by SIX which gives you a bigger slice. Of course, if you used a calculator work out $1 \div 6$ you would get 0·1666666 which is still useless.

It's much easier just to say each person gets *one sixth* of the cake, then cut the cake into six equal pieces. It's like magic because you will have multiplied the cake by 0·1666666 to make each piece. By gum, aren't you clever?

Even the poshest calculators can be rubbish

If you try to put $1 \div 6$ into different calculators, they might not even get the same answer!

If it's only got a little screen it might say that $1 \div 6 = 0 \cdot 1666666$

A posh calculator with a really long screen might say that $1 \div 6 =$

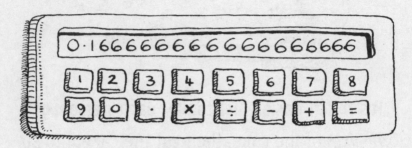

So, which one is right? The answer is that neither of them is exactly right! What if your calculator had a million digits? There are two problems here:
1 It would not fit in your pocket (unless you were wearing some very funny shaped trousers).
2 It still wouldn't be exactly right!

The trouble is that when the calculator works out $1 \div 6$, it keeps getting little bits left over, so it has to keep on dividing. These little bits get teenier and tinier but they never completely disappear. The sums go on for ever until the calculator runs out of space or it drops dead.

Five things to do with a dead calculator

1 Wrap it in silver paper and pretend it's a bar of chocolate.

mmm yum!

2 Hollow it out and use it as a very thin ice cube tray.

3 Stick it on your chest and pretend to be an android.

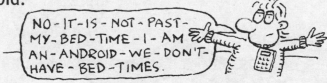

NO-IT-IS-NOT-PAST-MY-BED-TIME-I-AM AN-ANDROID-WE-DON'T-HAVE-BED-TIMES.

4 Pretend it's a mobile phone

5 Rent it out for ants to live in.

The thing about avoiding silly numbers like 0·1666666 and 0·1428571 is that it keeps your head nice and clear, and you'll need a clear head if ever you come across...

Urgum the Axeman!

Urgum has three sons called Ruff, Rack and Ruin, but he doesn't have a daughter called Ruth. In fact they are all *ruthless*.

Urgum also has eleven axes. He has promised that Ruff can have half of them, Rack can have a quarter of them and Ruin can have a sixth of them. How would you help them share the axes out? Remember, these guys won't be happy if you get it wrong!

(The calculator clot would start by working out $11 \div 6$ then he would tell Ruin he was going to get $1 \cdot 8333333$ axes. Ruin would then get confused and angry, and he'd start using the clot for chopping practice.)

The answer is quite tricky, but just in case it ever happens to you, here's what to do...

First you will need to ask Urgum if you can borrow one more axe. Urgum will smile sweetly at you and say, "Sure, but make sure you bring it back or else ... har, har, har!"

Put it with the other axes to make twelve axes altogether.

Now then, Ruff gets half of them which means you multiply the twelve axes by a half, which looks like this: $12 \times \frac{1}{2}$. This is the same as dividing 12 by 2, and the answer is 6. Ruff gets six axes ... but before Ruff takes them away, work out the others.

Rack gets a quarter of the axes (a quarter means the same as a fourth), so divide 12 by 4 and you'll see Rack gets three axes.

Ruin gets a sixth of the axes, so divide 12 by 6 and you'll see Ruin gets two axes.

"OK lads, help yourselves," you say.

Ruff takes *six*, Rack takes *three* and Ruin takes *two*.

Good grief! There's one axe left over. You quickly give it back to Urgum and run off as fast as you can.

Can you see how this puzzle works? If you add up the fractions each of the three boys has, this gives you $\frac{1}{2} + \frac{1}{4} + \frac{1}{6}$ ... which does not quite make 1. (It makes $\frac{11}{12}$. You can find out how to do sums like this in *The Secret of Sums*.)

When you borrow the extra axe, that means you have twelve axes which makes the sums easier, but because you only need to hand out eleven axes, that means you have one left over to give back.

The herd stood silent in the pre-dawn gloom. In the distance two points of light made their way down the hillside towards it. Chainsaw peered through the windscreen of the Dodge Sedan.

"This is as far as the track goes, Boss," he said. "I guess we're on foot from now."

"Kill the lights," snapped Blade. "Go on then, guys, you got beefs to shift."

The doors creaked open and a foot in a white leather shoe gingerly reached down to the ground.

"Feels kind of muddy," said Jimmy. "Where's the sidewalk?"

"There ain't no sidewalk, dummy," sneered Half-Smile. "This is the country, see?"

"I don't know about see, but I sure can smell," said Jimmy. "And what I can smell is stuck to my shoes."

"Stop beefin'," said Blade. "Besides, we don't even know what time we're supposed to block the line."

"I was thinking about that," said Weasel. "The train goes at 30 miles per hour, and we're five miles from the Fort, right?"

"That's what Doll said," said Blade.

"Well, if the train takes an hour to do 30 miles," said Weasel, "then it don't take nearly so long to do five miles."

"Six fives are 30," said Numbers.

"Right!" said Weasel. "So in one hour the train could do six lots of five miles."

"Why would it want to do that?" asked Blade.

"No, you don't get it," said Weasel. "I guess this train takes one sixth part of an hour to do the five miles."

"One sixth part of an hour – ten minutes!" said Numbers.

"There you are!" said Weasel. "The train will be passing here ten minutes before it's due to reach the Fort."

"Hey! It's due to reach the fort at 6:40," said Chainsaw.

"So ten minutes before that is 6:30," said Half-Smile. "That's when the train's due past!"

"Guess you're all feeling pretty smart with your little bitty maths, huh?" said Blade.

In the darkness of the car, the others all grinned smugly.

"Well, I make the time 6:27," said Blade. "And my little bitty maths says you've got three minutes to wade through whatever these beefs have left on the ground and get them out onto that line, so move it!"

THE SHORTEST CUT

It was early evening at the Last Chance Saloon. Riverboat Lil and Brett Shuffler had spent a long day playing cards, and as usual Lil had been winning. Brett had started with one thousand dollars but he had lost half of it.

"Aw shucks!" moaned Brett. "I'm not playing with you no more Lil. You already won five hundred of my dollars."

"But you might win them back!" said Lil.

"Forget it Lil," said Brett. "There's no way I'm losing my other five hundred."

Just then the saloon doors swung open. In staggered Will Nugget, the old gold prospector clutching a piece of yellowed paper.

"That's me done!" he announced. "I don't want to see another speck of gold as long as I live, so I'm selling up and getting out. This map shows where I buried my gold – and the price is one thousand bucks!"

"Hey!" whispered Lil to Brett. "He's been working those hills for months. There must be a fortune there."

"Right," agreed Brett. "But I only got five hundred dollars left."

"I've got five hundred too!" said Lil. "So together we've got one thousand, enough to buy the map. Do you want to be my partner?"

"Count me in!" said Brett.

Soon they were studying the map they had just bought.

To Find the Buried Gold

From Horseshoe Rock ride:
East 11 miles
North 4 miles
West 2 miles
South 8 miles
West 10 miles
North 3 miles
East 1 mile...
and get digging.

Horseshoe Rock

Scale
0 2 4 6
Miles

"We set off from Horseshoe Rock," said Lil.

"C'mon then!" said Brett and the two of them went outside. Brett leapt up onto his big horse while Lil carefully climbed onto her little pony.

It was sunset when they reached Horseshoe Rock,

and Lil was looking tired and sore. She slid down from her pony and leant against the big curved stone.

"Now we start following the directions," said Lil. "But first I need a little rest."

"Then I guess I'll have to wait for you," said Brett. "Although it seems mighty peculiar that you and me are working together. Usually you're cheating me out of my money at some card game or something."

"We could play a game now if you like," said Lil. "How about a race for the gold?"

"WHAT?" gasped Brett. He looked down at Lil's funny little pony. "Did I hear you right? You want to race me for the gold?"

"Why not?" said Lil. "I get bored of winning all those bar room games. Who knows, maybe I can ride well enough to beat you!"

Now there was one thing Brett was sure of.

Although Lil had a hundred ways of cheating him with cards or dice or coins, he could outride her any time he liked. He licked his lips and tried not to show how excited he was.

"Sure," he said. "If you want to race, who am I to disappoint a lady?"

"Then it's a deal," said Lil,  getting back on her pony. "Whoever reaches the gold first takes it all."

Brett couldn't believe his luck. There was no way he could lose.

"Who gets to hold the map?" asked Brett.

"You have it," said Lil. "I can remember the directions."

"Sure you can," smirked Brett. "And besides, you'll always have my hoof prints in front of you to follow!"

"You sound mighty sure about that," said Lil.

Brett was mighty sure.

"Come on then," he said. "What are we waiting for? Let's go! YEE-HAH!"

With a wave of his hat he charged off for the first eleven miles east. After a couple of minutes he looked behind. There was no sign of Lil.

"Yee-hah!" he cried again to himself. Just for once he *had* to win.

Brett rode and rode through the night keeping track of the miles and steering by the stars. Just as he was heading on the last mile east, the sun broke over the horizon. He could see something straight ahead of him! He rode up to find it was a small hole dug in the ground. In the bottom of the hole was a note.

> *I guess I can ride well enough* ✿
> *after all. Next time I see you, the*
> *drinks are on me.* *Love Lil* xx

"Oh no!" wailed Brett. "I lost the gold … AND I lost my last five hundred dollars!"

Back in the Last Chance Saloon, Lil and Will were laughing over breakfast.

"Just promise me one thing," said Lil. "Don't tell Brett there was never any gold."

"I won't!" chuckled Will. "Here's your five hundred dollars back, and here's Brett's five hundred dollars."

"That's not Brett's no more," chuckled Lil. "That's mine, but I'm gonna give you fifty dollars for coming in with the fake map."

Lil passed Will's share over and the old man tucked it under his hat.

"That was a great plan of yours, Lil," cackled Will. "Sure beats gold prospecting for a living!"

So how did Lil win the race?

When Lil planned the trick, she worked out a sneaky set of directions to put on the map. This is the route that Brett took.

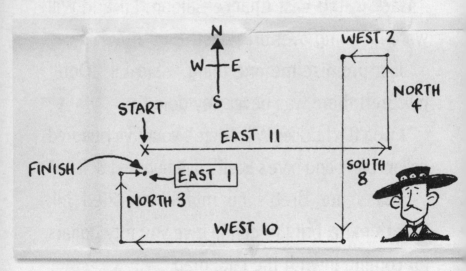

Even though Brett rode for a total of 39 miles, he actually finished 1 mile south of where he started! All Lil had to do was ride one mile south from the Horseshoe Rock and leave the note.

We don't even need a drawing to work out where Brett would end up. We just add up the miles that he travelled in each direction:

East: $11+1 = 12$ miles

West: $2+10 = 12$ miles

If Brett had just travelled a total of 12 miles east and 12 miles west, he'd have ended up where he started! So the east and west miles cancel out.

North: $4+3 = 7$ miles

South: 8 miles

He also travelled 7 miles north and 8 miles south, so when you put those together that means that he ended up 1 mile south from where he started which is what our diagram showed us!

WELL, SHOOT MA BOOT!

THE PSYCHIC BLOBS

This is a great trick to play on a friend who thinks you're mad for reading a maths book. Just say that the book has been infested by the five psychic blobs of numerical ectoplasm which are on the next page. When your friend tells you that you are totally nuts, you can go ahead and prove it!

1 Ask your friend to pick a secret number between one and 30 without telling you what it is.

2 Your friend now tells you which blobs have the secret number on.

3 Stare deeply at the psychic blobs, and explain that they are sending you a telepathic brain message.

4 You can tell your friend the secret number!

5 Have a safe chair and a glass of water ready. Your friend will need them to recover from the shock.

So how does it work? Look at the number immediately above the eyes on each blob. Add together these numbers on the blobs your friend picks. (You'll see that some numbers such as "4" only appear on one blob, but others appear on more – i.e. "23" appears on four of them.)

DANGEROUS CAKES

Over at Fogsworth Manor, the Duchess has organized a tea party, but there's an unexpected guest ... *Professor Fiendish!*

The Duchess had made four nice little cakes, but if you look at the plate you'll see there are *seven!*

The professor has played one of his diabolical tricks. Did you know that if you mince up lots of worms and slugs and boil them up with washing powder, you end up with a yellow lump that looks and smells like sponge cake? That's what the professor did, and now he's made three dangerous cakes and slipped them onto the plate with the others.

Luckily for us, each cake has a pattern of four little sweets on the top. The professor made sure that the pattern on his three cakes was different to the other four, so he could tell which cakes were his, even if they all got moved about.

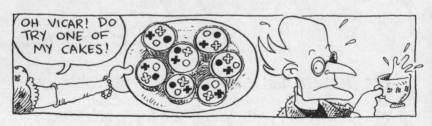

CanYOU tell what the dangerous cakes look like?

(If you turn the cakes around you'll find that the professor's three cakes match up.)

And here's another question for you…

How many patterns can you make with four different sweets on a cake? Each pattern must always look different from the others, even if the cakes get turned round. You could try drawing them all out!

While you're thinking about that, the professor has sneaked over to Primrose Poppet's display of hand-made bracelets.

HAR HAR! I'VE ADDED A FOURTH BRACELET BUT IT DOESN'T HAVE BEADS, THEY ARE MINIATURE STINK BOMBS!

Which bracelet has a different pattern to the others? At first you might think it's obvious, but the trouble with bracelets is that they can be flipped over as well as turned round. If you just have three different beads on a bracelet, there is only one pattern you can make. In fact the professor has no way of knowing for sure which bracelet is his!

Primrose has also made some bracelets with *four* different beads. This time the professor did manage to add an extra bracelet which is different to all the others.

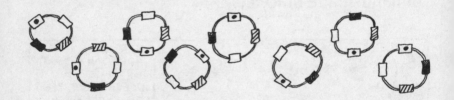

Can you spot it?

There are only three possible patterns with four beads, but how many different patterns could you make with five beads?

ANSWERS

This is what the three dangerous cakes all look like:

You can make six different patterns using four sweets.

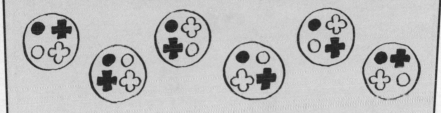

We can work this out using a funny little formula...
Number of cake patterns = (s-1)!
The little *s* is the number of sweets we put on the cake, so for four sweets, *s* = 4. Inside the bracket we get (4-1) = 3. We now know that the number of cake patterns is 3! but what does the ! mean? The ! sign is a *factorial* and it tells you to multiply the number by all the smaller numbers down to 1. Therefore 3! = 3×2×1 = 6.

Suppose we could put *five* different sweets around the top of each cake? The number of patterns would be (5-1)! = 4! = 4×3×2×1 = 24. So five sweets gives you 24 patterns, six sweets would give you 120 patterns, seven sweets would give you 720 patterns and eight sweets would give you ... 5,040 different patterns!

The only bracelet that doesn't match the others (even when it's flipped over) is this one:

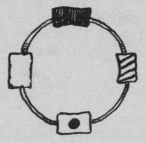

It's the only one that has the dotted bead opposite the black bead.

Because you can turn a bracelet over, each bracelet can make what looks like two different patterns. Therefore you only need half the number of bracelets to make all the possible patterns.

As the number of cake patterns is (s-1)! the number of bracelet patterns is $\frac{1}{2}$×(s-1)! However as the bracelets

use beads instead of sweets, we'll make the s into b. This means that the number of different bracelets with b different beads is
$$=\tfrac{1}{2}\times(b-1)!$$
So if we have five beads, the number of patterns

$$=\tfrac{1}{2}\times(5-1)!=\tfrac{1}{2}\times(4)!=\tfrac{1}{2}\times4\times3\times2\times1=\tfrac{1}{2}\times24=12$$

But the professor doesn't like formulas, so he's been getting his pet pig to help him!

AS YOU CAN SEE, HE'S BEEN WORKING OUT THE ANSWERS WITH REAL BEADS AND STRING!

DON'T GET DISGRUNTLED! HAR HAR!

MEANWHILE ON A RAILWAY SIDING...

With a soft clang the points on the track slid back into place.

"I guess that's Harri pulling the lever in the signal box," said Half-Smile. "So far so good, eh boss?"

"Come on," said Blade. "We got to get this train camouflaged before sun-up."

Blade watched as the men threw branches and leaves over the hijacked train. Somehow he felt uncomfortable – the job had been too easy. Just as planned, the train had stopped to allow the cattle over the line, the driver and stoker had even reversed it down the siding before they jumped off and disappeared up the hill. The gang had practically been given 15 million dollars. No, Blade didn't like it.

Chainsaw had found a small shutter in the side of one of the trucks.

"Boss," he called. "Look at this, it's loose."

Chainsaw hit the shutter with a stone. Suddenly it broke away and a stream of coins shot from the hole.

"Jackpot!" sniggered Weasel.

"Quick, you dopes," hissed Blade. "We need something to catch it in!"

"Come here, Weasel," said Chainsaw. "We need your pants."

Weasel protested, but the others had already grabbed him and were holding him against the truck so that the coins flew down the back of his trousers.

"It's c-cold!" stammered Weasel.

"Get his necktie and tie it round his legs," said Blade. "We don't want the dough falling out at the bottom."

Gradually the flow of coins trickled to a stop.

"Hey, Weasel, you shrinking?" asked Jimmy.

"Not shrinking, he's sinking!" said Chainsaw.

Sure enough, Weasel, weighed down by his bulging trousers, was gradually slipping into the ground.

"Grab him!" ordered Blade. "There must be a fortune in his pants."

"Two thousand seven hundred and fourteen dollars," said Numbers. "I counted."

"We better start shifting this stuff," said Blade. "Carry him back to the car."

As they struggled past the roaming cows, Weasel

managed to gasp: "Hey Numbers, how many 2,714s are there in 15 million?"

"Five-five-two-six," said Numbers, "and a little bit."

"For the last time," snapped Blade, "will you guys quit with the maths?"

"But boss," moaned Weasel, "we got to think of a better system than this. Otherwise you guys will be lugging me across this cow field five and a half thousand times."

"That's a lot of times," said Jimmy.

"I know how we could make it less times," said Chainsaw.

"How's that?" said Blade.

"We could get Weasel some bigger pants."

THE TROLL'S CHALLENGE AND THE DEADLY ISLANDS OF DOOM!

It's not your fault. It could have happened to anyone. There you were, lightly skipping through Fairyglen Forest picking buttercups and daisies when you tripped on an old gnarled tree root and fell down a pit into the caverns of Neverworld.

You look around and see the walls are covered in strange symbols, then you hear a muttered curse.

"No! It can't be done!" shrieks a voice.

"Then you must die!" cackles a mystic troll.

"Please ... one more chance!" begs the voice.

It turns out that many years ago a peasant fell down the same pit and was caught by the troll.

169

"Let me go," said the peasant, but the troll first set him a challenge.

"Choose one of the symbols on the wall," the troll had said, then he passed over a long piece of dirty string. *"You must arrange the string to make the same pattern as the symbol, but you may not double the string back on itself at any point."*

This is the symbol the peasant had chosen:

So, is the puzzle impossible, or is the peasant just a bit slow?

These puzzles are good fun (unless you're being threatened by a mystic troll).

Instead of using string, you can draw out the diagram but there are two rules:

- you must not take your pen off the paper.
- you must not go over any line twice.

This diagram is called the envelope puzzle because it looks like the back of a closed envelope:

Can you draw this diagram without breaking the rules?

The answer is NO.

Note: If some big-head tells you he can do it, then he's cheating. The only way he can do it is either by folding the paper over, or by rubbing a line out afterwards or by having a wonky pen that doesn't write sometimes. Remember that people like that usually don't have any friends and tend to get their hands stuck in vending machines.

Now then, look at this open envelope…

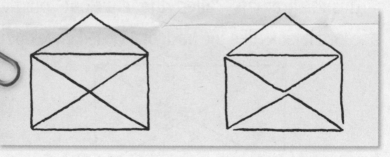

By starting at one of the bottom corners and finishing at the other, you can draw it!

How can you tell if you're going to be able to draw a pattern without taking your pen off the paper?

The secret is to look at all the places where lines join up. (These are called nodes.)

The closed envelope has five nodes. Four of them are the four corners, each of which has three lines going to it. There is also a fifth node in the middle which has four lines going to it.

You ignore any nodes which have an even number of lines going to them, so for the closed envelope you can ignore the node in the middle.

What you need to know is how many nodes have an odd number of lines going to them, i.e. 1, 3, 5, etc.

• The closed envelope has *four* odd nodes (each with three lines).

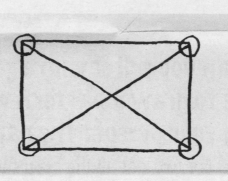

If you look at the open envelope, remember you can ignore any nodes with an even number of lines, so:

- The open envelope has *two* odd nodes.

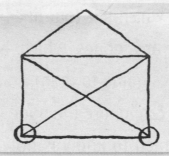

Now get this: You can draw any picture without taking your pen off the paper if there are two odd nodes or less.

Special node notes

• If a picture has no odd nodes, you can start and finish where you like.

• If there are two odd nodes, you have to start at one and finish at the other.

In other words, you could draw this picture if you wanted:

It has only two odd nodes.
(Can you find them?)

Before we go back to Neverworld, there are two more things to know.

1 If you have a line sticking out, then the end of the line is counted as an odd node. Look at these:

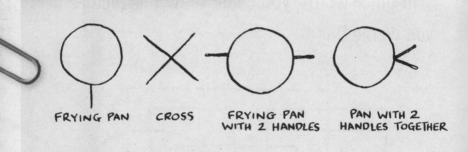

FRYING PAN   CROSS   FRYING PAN WITH 2 HANDLES   PAN WITH 2 HANDLES TOGETHER

The first frying-pan has two odd nodes, so you should be able to draw it without taking your pen off the paper.

The cross has four odd nodes so you will have to take your pen off the paper. As for the pans with two handles ... work it out yourself, it's dead easy.

2 You can only have an even number of odd nodes! That's because a line always has to have two ends. Try it ... scribble a few lines together then count up the odd nodes. There *has* to be an even number.

So back to Neverworld ... look at the symbol the peasant chose. How many odd nodes are there?

But now the troll has turned to you.

"Pick a symbol," he snarls horribly.

"Tum tee tum," you say casually, just to get him all wound up.

Here are the symbols. Which one should you pick if you want to escape?

Of course being dead clever you pick the right symbol and get out of Neverworld, but then as you speed away from Fairyglen Forest you take a wrong turn and find yourself in…

The foul city of Fastbuck

The city is built around Lake Nasty. All the filth has been poured into the lake, and all the radioactive and biological waste has been piled up on three islands called Yuk, Ug and Poo.

The islands are linked to the mainland and each other by eight bridges as you can see on this map:

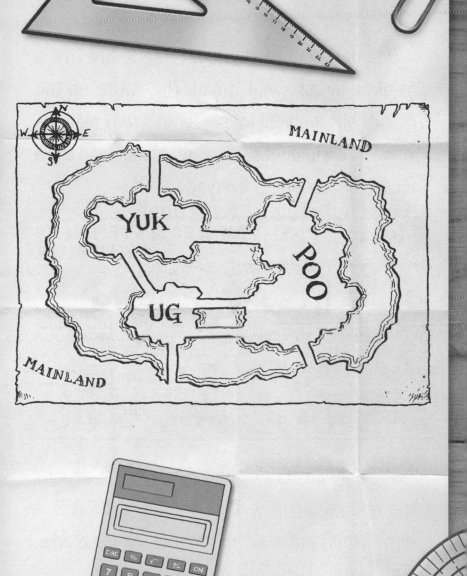

179

One dark and smelly night, the entire city is awoken by an awful groan. The waste on the islands has mutated into hideous Crud creatures who will bring disease, famine, plague and extra French homework to everyone.

The only way to stop this disaster is to destroy the bridges, and the best person to do it is a brave, calm and cool *Murderous Maths* reader. (That's you.) These are your instructions:

- You will be given a giant steamroller.
- To destroy a bridge you must drive the steamroller right across it.
- Once you've driven over a bridge, it is destroyed so you can't cross it again.
- You must start and finish on the mainland. (Of course you can go back to the mainland in between, too.)
- Don't get a parking ticket.

Can you work out which way to go to cross all eight bridges just once each?

After hours of driving around in your giant steamroller, you will realize it's impossible. There will always be at least one bridge left standing, and you might end up stuck on one of the islands as well.

Just as you are about to give up, a batty billionaire suddenly says, "Would it help if I quickly built an extra bridge?"

The citizens of Fastbuck all groan.

"We want the bridges knocked down," says the mayor, "we don't want any more!"

"But if the steamroller went over my new bridge, it would get knocked down just like the others…" explains the billionaire.

The crowd all start to mutter "so what's the point of that?" but then…

Flash! A giant light bulb suddenly comes on in your brain as you realize this is the same sort of challenge as the one you faced in Neverworld.

If you are brainy then stop reading! Can you work out how an extra bridge could help you?

Look at the map of the bridges. You have to go over each one once with no doubling back, just like when you were drawing the troll's mystic symbols.

To make it clearer, we'll draw a diagram of all the bridges like this:

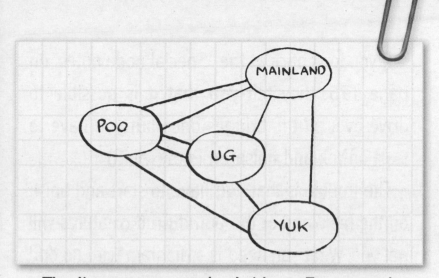

The lines represent the bridges. For example, there are two bridges going from the mainland to Poo, so there are two lines on the diagram. There is only one bridge going from Ug to Yuk, so the diagram only has one line, and so on.

Now you can solve the puzzle because the islands (and the mainland) are nodes! You'll see there is an even number of lines going to Ug and the mainland and an odd number of lines going to Yuk and Poo. This means that you have two odd nodes!

If you look back at the "Special node notes" on page 175, they tell you that it is possible to drive over all the bridges once but you have to start on Yuk and finish on Poo. (Try it!)

The trouble is that you have to start and finish on the mainland or the putrid Crud creatures will get you. What you need is a diagram with no odd nodes, then you can start and finish where you like. This is where the batty billionaire comes in.

Suppose you put an extra bridge between Yuk and Poo?

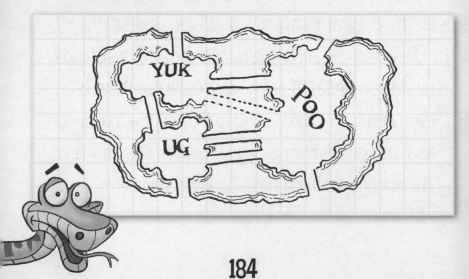

You haven't got any odd nodes now, so you can start and finish wherever you like. That means you can start and finish on the mainland, which is what you want.

Try it out! You should now be able to destroy all the bridges (including the new one) and get back to the mainland in time to save the city from the creatures of Crud.

THE GREAT RHUN OF JEPATTI
AND HIS SQUARES OF MYSTERY!

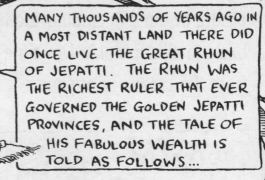

MANY THOUSANDS OF YEARS AGO IN A MOST DISTANT LAND THERE DID ONCE LIVE THE GREAT RHUN OF JEPATTI. THE RHUN WAS THE RICHEST RULER THAT EVER GOVERNED THE GOLDEN JEPATTI PROVINCES, AND THE TALE OF HIS FABULOUS WEALTH IS TOLD AS FOLLOWS...

The palace of the Rhun had a tower of 60 chambers for his sons to live in. However the Rhun had only been blessed with 59 heirs before his wife decided the ancient washing machine couldn't cope and announced, "No more!"

And so it came to be that in one chamber dwelt not a son but the Ghinji, a loathsome dragon-toad and devourer of human flesh.

Many families would visit the Rhun and present their finest daughters for marriage. On payment of a sack of gold coins, the girl

would choose a chamber, and enter therein. The Rhun promised that on finding a son, the couple would marry the very next day.

Yet on finding the Ghinji, the door would be closed and the girl left to her fate.

And here, dear reader, lies the mystery: no girl ever did ever marry a son for they all came to rest in the belly of the Ghinji.

BURP!

BOTHER!

Wow! Gruesome story, eh? But why did every girl pick the room with the Ghinji? The Rhun's secret relies on a mystery square of numbers.

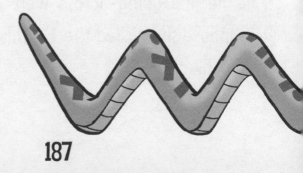

The doors to the rooms were numbered from 1 to 60. Each girl was shown a great tablet of stone with this mystery square carved on it.

The girl had to choose one number from the square and then a circle was drawn round it. All the numbers in the column above and below were crossed out, and all the numbers in the same line going across were crossed out too. Suppose she picked the 13, this is what it would look like:

She then had to choose any other number that was not crossed out. Another circle went round it and the numbers in the same row and column were crossed off. Here she's picked number 1.

Following the same rules, she had to choose two more numbers. (Here she picked 10 on the bottom line, then the 8). That would leave one last number, which also got a circle round it. Here it's the 7.

The circled numbers were added up, so here she'd get 8+7+13+1+10 = 39. This was the room number she had to go to.

Here's the spooky bit – it never mattered which numbers were chosen – the answer was always 39!

Suppose the girl picked the 6 to start with, and after that the 7 on the bottom line, then the 4, then the 12 on the left-hand side…

…the only number left is the 10 at the top. Add them up: 6+7+4+12+10 = 39!

No wonder the Rhun was so rich. All he had to do was make sure his Ghinji was behind door 39 and ready for lunch.

How to make your own mystery squares

You can make a mystery square to produce any number you like. (If you have 59 sons and a Ghinji, then you could make a load of money too!)

All you need is a grid with 25 squares and a friend called Patsy to try it out.

GRID   PEN   PATSY

Ask Patsy to choose any number over 20 (but don't let her go too high or it gets harder).

Let's say Patsy chooses 27.

First of all we'll make the *simple version* of the mystery square:

1 Put 0, 1, 2, 3, 4 along the bottom line. OK, now here comes the bit which needs practice…

2 Take away ten from Patsy's number. If Patsy said 27, the answer is 27-10 = 17.

3 Split this number into four smaller numbers. For 17 we'll choose 2, 3, 5, 7 because 2+3+5+7 = 17. (You can pick any four numbers as long as they are different and they add up to 17.) Try to do this in your head.

4 Put these four numbers in the column above the zero. They can be in any order.

7				
2				
5				
3				
0	1	2	3	4

5 Fill in the rows of the grid by "counting up" from the edge. So for the top row we start with 7, then count up 8, 9, 10, and 11.

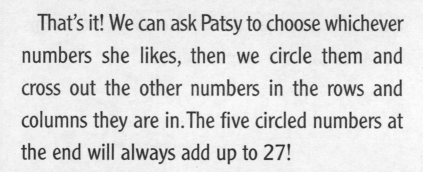

That's it! We can ask Patsy to choose whichever numbers she likes, then we circle them and cross out the other numbers in the rows and columns they are in. The five circled numbers at the end will always add up to 27!

Now that you can make the simple version of the mystery square, it's time to try the Great Rhun's clever version. At the moment we've got 0, 1, 2, 3, 4 on the bottom row and if Patsy sees that she might guess how we made the square. We don't want that, do we? No we don't, so this is how to mix the numbers up…

196

1 Before we start we give Patsy the pen and she can put a zero in *any square she likes*!

2 As soon as she has put in the zero, we fill in the numbers 1, 2, 3, 4 on the same line in any order.

3 Let's suppose that this time Patsy picked 41 to be her number. First we take away 10 to leave 31.

4 Now we split 31 into four different numbers. 11+5+7+8 = 31 so the four numbers we'll use will be 11, 5, 7 and 8.

5 Fill these numbers in any order above and below the 0.

			7	
			8	
3	2	4	0	1
			5	
			11	

6 Fill in the other numbers by counting up as before, but put them in above (or below) the 1, 2, 3 and 4 in order.

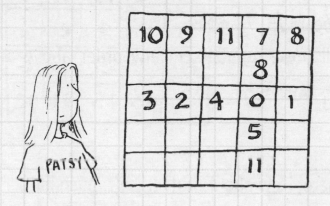

10	9	11	7	8
			8	
3	2	4	0	1
			5	
			11	

On the top line here, we had a 7 above the 0. This means the number above the 1 is 8, the number above the 2 is 9, the number above the 3 is 10 and the number above the 4 is 11.

Here's the completed mystery square for number 41:

PATSY

10	9	11	7	8
11	10	12	8	9
3	2	4	0	1
8	7	9	5	6
14	13	15	11	12

There! How about that? It isn't much harder to do, but it is a lot more complicated for Patsy to figure out. She'll also be flummoxed because she could choose which square had the first "0" on it!

There's one extra brilliant thing about this trick. It even works on people who aren't called Patsy!

Benni the waiter hovered nervously behind the counter. He recalled the shady figures seated around table 12 all too well. Last time they had visited, there had been a massive fight, and he had only just replaced all the broken plates and glasses. Mind you, thought Benni, this time they were sitting with a classy dame so maybe they were on their best behaviour.

"There it is, Doll," said Blade. "One banker's cheque for all the bail money, plus interest."

"You did well, guys," said Dolly Snowlips. She folded the cheque and slipped it into a secret trapdoor in one of her high heeled shoes.

"It was tough work getting all that dough to the bank," said Half-Smile.

"You don't say," sniggered Dolly. "By the way, Weasel, what's the matter with your pants? I seen smaller circus tents."

"Don't ask," replied Weasel, who could just about see over the belt buckle of his gargantuan trousers.

"So now we've paid our dues," said Blade, "we'll be on our way."

"What's the hurry, boys?" asked Dolly. "Don't you want to know who this cheque is going to?"

The gang looked at each other uneasily, but then the door opened.

"Evening, boys," said the prison governor. "Long time no see!"

"It's the Beak!" gasped Chainsaw.

"Relax," said Blade. "We're all paid up. He can't touch us."

"You paid, huh?" sneered the governor. "Is that right, Doll?"

"Not a cent, honey," said Dolly who had gone over to close the door behind him.

"Is that a fact?" said the governor. From inside his coat he pulled a nine-barrelled Dawson-Roach 25-bore multi-target self-loading hydraulic action scatter cannon.

"Oh mother of mercy!" whimpered Half-Smile.

"Eight-fifty slugs a second," muttered Numbers. "Nice piece."

"The way me and Doll see it," drawled the governor, "you seven saps escaped jail then robbed the Knocks Express. Boy, are they going to give you a welcome at Grimstate."

"You mean you double-crossed us?" cursed Blade. "I knew it! That job was far too easy, what with the engine crew parking the train, then running away just like that..."

"Harri's brothers," said Dolly. "We didn't leave nothing to chance."

"Face it, guys," sighed Blade. "Doll and the Beak have fixed us real good."

"You always were a loser, Blade," said Dolly. "When you're back inside taking the rap, me and the Beak will be taking a long flight."

"Just tell us one thing," said Weasel. "Who chucked the dynamite into the cell?"

The governor smiled.

"I got to take credit for that. You guys were getting it too easy in there, I figured you needed encouraging to get out more."

Dolly called over to Benni.

"Hey, you behind the counter. Pull down the

shutters. This place closes early tonight."

Benni hurriedly obliged.

"And when you've done that, go call the Feds," said the governor. "In the meantime, just one of you has to twitch and *kapow*, you're all bolognese sauce."

Whump!

The door suddenly burst off its hinges and landed on the floor, completely sandwiching Dolly and the governor underneath. A massive figure staggered inside.

"It's Porky!" gasped the gang.

"Am I too late?" puffed the figure who had come to a halt standing on the door. "I just saw the place was closing up early, so I was running to get a last order in, but the door was shut and when I'm moving fast I don't stop so good and gee, I'm sure sorry about the damage, Benni."

From under the door came some muffled groans.

"Porky," ordered Blade. "Don't move your feet from that spot. Jimmy, shove him a chair. The rest of you guys, move the table right up to him. Benni, you go and get three of everything on that menu and put it out in front of my biggest and bestest buddy here."

"But boss," said Porky, "I can't afford three of everything."

"That's OK, big man," grinned Blade. "Tonight you're having dinner on the Beak."

Three hours later the party was still going on.

"…and you know what the best part is," said Blade. "The Beak came in here to arrest a gang of seven men – but he didn't realize there was only six of us here!"

If only the governor had stopped to count them up the story would have ended very differently.

It just goes to show, even the most basic maths can be murderous.

INDEX